YOUR KNOWLEDGE HAS VALUE

- We will publish your bachelor's and
  master's thesis, essays and papers

- Your own eBook and book -
  sold worldwide in all relevant shops

- Earn money with each sale

Upload your text at www.GRIN.com
and publish for free

Effect of physical soil and water conservation measures on physico-chemical properties of soil

Fichago Selman

Bibliographic information published by the German National Library:

The German National Library lists this publication in the National Bibliography; detailed bibliographic data are available on the Internet at http://dnb.dnb.de.

ISBN: 9783346766342

This book is also available as an ebook.

© GRIN Publishing GmbH
Nymphenburger Straße 86
80636 München

Print and binding: Books on Demand GmbH, Norderstedt, Germany
Printed on acid-free paper from responsible sources.

GRIN web shop: https://www.grin.com/document/1299382

HAWASSA UNIVERSITY

WONDO GENET COLLEGE OF FORESTRY AND NATURAL RESOURCE SCHOOL OF FORESTRY PROGRAM URBAN FORESTRY & GREENING

A SENIOR SEMINAR

ON

EFFECT OF PHYSICAL SOIL AND WATER CONSERVATION MEASURES ON PHYSICO-CHEMICAL PROPERTIES OF SOIL

A SENIOR SEMINAR

BY

FICHAGO SELMAN

HAWASSA

ETHIOPIA

November 2022

Table of content

Contents   **page**

List of tables

Abbreviation

MoA-   Ministry of aricultur

OC-   Organic carbon

SLMP-   Sustainable land management program

SOM-   Soil organic matter

SOC-   Soil organic carbon

SWC-   Soil and water conservation

TN-   total Nitrogen

1. Introduction

Soil degradation is deterioration of the physical, chemical and biological or economic properties of soil (Maitima and Olson, 2001). It has a negative impact on agricultural economy and the natural environment (Taddese, 2001, (Keno and Suryabhagavan, 2014). Human activity, such as conversion of forests to agricultural land, increased cultivation of marginal land, overgrazing, and subsistence agriculture practiced on marginal lands with steep gradients; accelerate soil erosion (Lal, R., 1996). Increasing population pressure has lead people to use marginal lands for cultivation and grazing and severely overexploit the natural resource base of the region (Hurni, 1993; Shiferaw and Holden, 1998). The exploitation of these natural resources is inextricably linked to securing food and livelihoods (Dubale, 2001; Adimassu et al., 2012). This has resulted in accelerated soil erosion, deforestation, water resource depletion and environmental degradation (Desta, 2000; Berry, 2003; Mitiku et al., 2006; Nyssen et al., 2008; Change, 2014).

Land degradation and its related decline in the productivity potential of agricultural land are challenging the economic and social well-being of the current and future generations on earth (Haregeweyn et al. 2012). Soil erosion is the main cause of land degradation and a leading factor contributing to poor agricultural development in developing countries (Gemechu 2016). Currently, soil resources are the main sources of livelihoods for most people of the world, such human exploitation being the foremost factor for soil degradation (Molla and Sisheber, 2017). The sorting action of erosion removes large proportions of the clay and humus from soil, leaving behind the less productive coarse sand, gravel, and in some case even stones, impairing the quality of the remaining topsoil (Troeh F., 1980). The removal of this organic matter affects soil properties including texture, structure, nutrient availability and biological activity and makes soil more susceptible to further erosion as its aggregates becomes less stable thus, negatively affecting crop production (Alemayehu, F. and Dubale, P. ,2006)

In Ethiopia, measurements from experimental plots and micro-watersheds showed the annual soil loss from croplands is about 42 t· ha–1·year–1 (Wagayehu et al,2003).As a consequence, the productive capacity of Ethiopia's highland soils is being reduced at an annual rate of 2% - 3%, which certainly contributes to a higher vulnerability to famine (Wagayehu et al ,2003) . The slope steepness, long cultivation history with outdated technology, and overgrazing make soil erosion more severe in Ethiopia (Nyssen et al., 2004). It has been identified as a major threat to the national economy (Hurni, 1993) and among the main challenges influencing the sustainability of agriculture (Molla and Sisheber, 2017). As a result, two-thirds of the population of Ethiopia has been affected by soil erosion mainly associated with the conversion of forest to agricultural land (Hurni *et al.*, 2015).

To solve this problem, the first initiative of soil and water conservation (SWC) investment began during the outbreak of the 1973/74 drought in Ethiopia(Berhe ,1996).The main intent of the initiatives was to minimize erosion, restore soil fertility, rehabilitate degraded land, and increase agricultural productivity (Mekuria *et al.*,2007). Since the 1980s, various national SWC efforts have been undertaken with the financial support of international donors and mass mobilization of rural communities (Holden *et al.* 2001). The largest SWC investment made in the country was during the Derg Regime in which more than 1billion US dollars per year were invested during 1974–1991 (Rahmato 1994). Since the 1990s, the implementation of SWC measures has been an integral part of agricultural extension packages (Bewket and Sterk 2002). Since the overthrow of the Derg Regime in 1991, investments in SWC in Ethiopia have continued. For example, more than 500million US dollar has been invested in the Productive Safety Net Programmed since 2005 in which the majority of the money was allocated to SWC activities (Gilligan *et al.*, 2009; Andersson *et al.* 2011). Moreover, huge financial resources have been invested in Sustainable Land Management Program (SLMP) since 2008 with the support of World Bank and Global Environmental Facility (Nedassa *et al.*, 2011) and MERET (Managing Environmental Resources to Enable Transitions to sustainable livelihoods) project since 2003 with the financial support of WFP (Zeleke *et al.*, 2014).

The effectiveness of SWC measures undertaken in Ethiopia was evaluated by several authors (Haregeweyn *et al.* 2016; Nyssen *et al.* 2010). According to (Dagnew *et al.*, 2015), these evaluation reports showed that there is no consent on the effectiveness of SWC interventions among them. Some argue that SWC contributes for reduction in runoff and sediment loss (Mekuriaw, 2017) and increased soil moisture conservation (Haregeweyn *et al.*, 2015; Nyssen et al. 2010). On the other side, it is reported that SWC structures were not effective in reducing soil erosion (Bewket and Sterk, 2002) and had not resulted in decreasing sediment concentrations (Temesgen *et al.* 2012). For example, construction of SWC practices such as soil and stone bunds reduced crop yield up to 7% for the first few years in Ethiopia (Adimassu *et al.* 2014; Kassie *et al.* 2011; Kato *et al.* 2011; Shiferaw and Holden, 1999). On the contrary, stone and soil bunds increased crop yield up to 10% in the Tigray region of

Ethiopia (Nyssen *et al.* 2007; Vancampenhout *et al.* 2006; Gebremedhin *et al.* 1999). This shows that the results regarding the economic viability of SWC practices are inconsistent and site-specific.

In the case of on soil property, reports show that most of SWC measures   have important implications for improving soil fertility (Belayneh *et al.* 2019). Tesfaye and Fanuel, 2019) also reported that SWC practices have positive effect in improving soil fertilities and crop yield in Southern Ethiopia. In other site , K.Wolka *et al.* 2011) reported that Most soil parameters were not significantly different in cropland with level soil bund and Stone Bund compared to non-terraced .This indicates that there is a gap in the evaluation of the impacts of SWC interventions.

Although many studies confirmed the positive impacts of SWC practices on soil physicochemical properties and crop yields, farmers frequently destruct SWCPs constructed on their fields, claiming that the practices didn't show a positive impact/effect other than occupying their farmlands. Such claims need investigation and measured data to design alternative land management strategies.

Therefore, the main objective of this review is to discuss the effect of physical soil and water conservation practices on physicochemical properties and fertility status of soil

2.   The states of Soil and Water Conservation in Ethiopia

2.1. Soil and water Conservation structures

Soil and water conservation structures include all mechanical or structural measures that control the velocity of surface runoff and thus minimize soil erosion and retain water where it is needed(Bancy M. Mati). They usually consist of engineering works involving physical structures, made of earth, stones, masonry, brushwood or other material for the construction of earthworks such as terraces, check dams, and water diversions, which reduce the effects of slope length and angle. SWC structures can be designed either to conserve water or to safely discharge it away. They supplement agronomic or vegetative measures but do not substitute for them. Suitability of SWC structure depends on Climate and the need to retain or discharge the runoff, Farm sizes and Soil characteristics

The prevention of erosion on cultivated land and other areas depends essentially on the reduction of soil detachment and runoff on the maintenance of adequate vegetation ground cover (Fitsum S., 2002). Soil conservation involves the various methods used to reduce soil erosion to prevent depletion of soil nutrients and soil moisture and to enrich the nutrients status of the soil. The conservation techniques include terracing and other.

2.2. Physical Soil Conservation

A physical soil conservation practice is applicable of soil management using knowledge or art with the goal protection of soil resource form exploitation. In addition, among those different applications, different structure applied in different farmlands .However, these conservation applications depend on climate, soil type, vegetation cover and level of economy.

3. Importance of soil and water conservation for sustainable agriculture

Effective soil and water conservation the major steps for sustainable agriculture. Sustainable agriculture can be practices by reducing negative measures and technology (Fitsum S. ,2002).To bring sustainability is on the basis of soil and water conservation, natural resource management, land management, integrated pest management using new technology, economic viability and food security on household level.

SWC practices in highland areas can foster the production of various kinds of ecosystem services that have both upstream and downstream benefits. With consciences with the farmer if proper implementation of techniques that maintain or restore the capacity of soil to retain water with the inorganic nutrients and organic matter increases.

Many studies in Ethiopia confirmed the positive impacts of soil and water conservation practices (SWCPs) on soil physicochemical properties and crop yields. For instance, soil conservation practices tested in Simada district, northwest Ethiopia, significantly improved the soil physicochemical properties (Mihrete, G.2014), Similarly, significantly lower mean bulk density was found in fields treated with SWCPs than the untreated fields in Adaa Berga district, western Ethiopia(Abay,C., 2016). Other studies conducted in Ethiopia and other countries also verified the positive impacts of SWCPs on soil physic-chemical properties and crop yields.

There are there types of soil properties, which are the physical properties (texture, structure, density, porosity, color, moisture content, water holding capacity etc) ,chemical properties (nutrient content, PH, CEC, electrical conductivity, redox potential etc.) and Biological properties (macro and microorganisms).

4. Impacts of SWC Practices on soil physical properties

4.1. Bulk density (BD)

Previous researches showed that SWC structures affect the bulk density of soil. The BD Soil treated with SWC structures is minimal compared with adjacent non-treated soil. S. Hishe et al(2017) reported that the highest mean bulk density in Middle Silluh Vally (Tigray) was recorded in non-terraced farm land (1.65 g/cm$^{3)}$ followed by terraced farm land (1.6 g/cm^3). The same result was found by Tesfaye and Fanuel (2019) in Southern Ethiopia. According to their result, Soil bulk density (BD) was affected by soil conservation practices. It ranges from 0.96 g/cm^3 (physical SWC for 5 years) to 1.10 g/cm^3 (non-conserved crop land) .K. Wolka et al. (2011) also reported that the effect of SWC on the mean soil bulk density was found to be minimal and slightly lower values were observed in conserved plots. Similar results were reported in different part of Ethiopia by (Yiferu *et al* 2018 , Gebiresilassie *et al*. 2013; Dulo *et al*. 2017; Solomon *et al*. 2017, Demelash and Stahr 2010,) . This could be due to the subsequent effects of reduced soil loss and crop residue through erosion, presence of higher organic matter resulted from conservation measures and decay of plant residues. A relatively higher bulk density in non-conserved plots could be related with washing out of fine organic matter rich soils by erosion and thereby exposed slightly heavier soil particulates.

4.2. The soil texture

The soil texture of the soil was assessed based on proportion of three mineral particles, sand, silt and clay.

4.2.1. Sand content

In Studies reviewed in this paper showed that the sand texture was relatively highest mean value in the non-conserved areas than conserved areas. On contrary, the lowest sand content was observed in conserved areas. S. Hishe *et al* (2017) reported that the sand content of non-terraced land was the highest mean (70%) and the terraced land was the lowest mean (62%).

4.2.2. Silt content

Since SWC structures reduce loss of soil particle, it increases the silt content of the soil. For example research conducted Tigray region provide evidence of silt content of the soil was highest in conserved land than non-conserved lands (S. Hishe *et al* 2017) that the silt content of the soil of terraced farm land was 13.5 % and non-terraced farm land silt content was 9.5%. The same result reported in Southern Ethiopia, that the silt fractions showed significant difference (P < 0.05) in croplands under terrace when compared with adjacent-non terraced croplands (K. Wolka *et al*. 2011)

4.2.3. Clay content

Similar to silt content, the clay content of the soil also affected by SWC structures. In Tigray , The clay content of terraced farm land was 26.5% and clay content of non-terraced farm land was 20.5% (S. Hishe *et al* ,2017).The same result was recorded by K. Wolka *et al*. (2011) , The clay fractions showed significant difference (P < 0.05) in croplands with level soil bund when compared with adjacent-non-terraced croplands. Belayneh et al, 2019 also reported that clay content experiencing a mean value of 67.8% and 60.5% in conserved and non-conserved soil respectively in Northern Ethiopia.

5.   Impacts of SWC Practices on soil chemical properties

5.1.   PH

Soil pH affects solubility and availability of nutrients in the soil (Mc. Cauley *et al*, 2003). The exchange reactions between soil solution and the soil particles surfaces are the main regulators of soil pH (Coleman *et al*, 1989).Soil and water conservation practices affect PH value of soil. For example, M.Guadie *et al*. (2019) reported that Soil pH showed a statistically significant difference (p≤0.05) between the treated and untreated fields in Northern Ethiopia. Belynch *et al* (2019) also reported that Soil pH showed slightly higher mean values in conserved plots. The analysis of variance result show that soil pH was not statistically significantly affected by conservation practices .Similar results were reported by Challa *et al*. (2016) and Husen *et al*. (2017) in the central highland of Ethiopia. This might be due to the effect of soluble bases and organic matter removal through sheet erosion from the control fields due to the absence of SWCPs, and due to the low base saturation percentage and low sediment organic matter (SOM) content and high pH value in the sediment accumulation zone behind the SWCPs of the treated fields. Relatively higher soil acidity in non-conserved plots may be related with high rainfall, associated with leaching and removal of important soil nutrients. High amount of rain water leaches soluble bases and consequently contributes to soil acidity.

5.2.   Soil organic carbon (SOC)

The accumulation of SOC is one of the main soil forming processes and is determined by physical, chemical, biological and anthropogenic factors with complex interaction (Gaiser & Stahr, 2013). Previous studies in Ethiopia reported that the SOC content of the soil is affected by SWC measures. SOC showed a statistically significant difference ($p < 0.05$) between the treated and untreated fields in Northern Ethiopia (M.Guadie *et al* 2019).

In Southern Ethiopia, Gola watershed it is reported that relatively high value of organic carbon content in treated land than non-treated land. Both of the mean values of organic carbon 1.18 for treated and 1.10 for untreated farm fields are found were not significant (H.Gankiso 2017). S.Hishe *et al.* also reported that a small difference in SOC concentration was found between terraced and non-terraced land. The SOC has a direct relationship with the vegetation cover and conservation measures application and inversely related with intensive human and livestock interference. Landscapes with different management had a significant effect on SOC. The same result was recorded in Southern Ethiopia (Wolaita Zone) that SWC practices influenced soil organic carbon (OC) of farmlands positively. The mean value of soil OC range between 1.34 and 1.74% in which integrated SWC established for 5 years had the highest value and the minimum was obtained on non-conserved land (Tesfaye and Fanuel , 2019). Overall, it was noted that the longer the age of SWC practices and its integration with biological measures, the positive is its impact on soil OC of cultivated lands. This might show that SWC practices have a positive role in improving soil OC. As a result, supporting terracing with vegetation could result in high SOC and SOM due to high biomass return, which contributes to symbiotic fixation and soil erosion reduction.

5.3.   Total Nitrogen (TN)

Total Nitrogen (TN) is the sum of all form of Nitrogen (ammonia, organic and reduced Nitrogen) and Nitrate-Nitrite in soil. The content of total nitrogen of the soil affected by SWC practices. S.Hishe *et al*, (2017) reported that the TN content had statistically significant difference at $p<0.05$ between terraced and non- terraced hillside in Northern Ethiopia. In finding of North Western Ethiopia, TN showed a statistically significant difference ($p \leq 0.05$)

between the treated and non-treated fields. The treated fields showed higher TN values than the untreated fields, which could be associated with the implementation of SWCPs that maintain soil fertility by decreasing the removal of SOC and TN through soil erosion. This finding is in line with Ademe *et al.*, 2017 and Mulugeta *et al.*, 2010, who found that higher TN content was recorded in treated fields compared with untreated fields in southern Ethiopia and northwest Ethiopia, respectively. In contrary H. Gankiso (2017) reported that the average total nitrogen content for both treated and untreated farm fields is rated low that it's not shows a significant difference. This low value of total nitrogen could be due to the fact that the area is moist kola and this may cause leaching of nitrogen in the soil. The other reason might be related to management problems in improving the nitrogen content by growing leguminous plants that fix nitrogen from the air through the nodules of their roots.

5.4. Available Phosphorus (Av-P)

Available Phosphorus (Av-P) showed a statistically significant difference between the treated and non-treated fields in North western Ethiopia (M.Guadie *et al* ,2019) . Low Av-P from non-treated fields was due to continuous cultivation without SWCPs, extractive crops biomass harvest, and soil erosion, as indicated by the findings of Dejene.T (2017) and Ademe *et al* (2017) in eastern and southern Ethiopia, respectively

The average available phosphorus high values of treated farm fields and low untreated farm fields with mean value 19.77and 18.11 are recorded in both management fields in Gola water shed, southern Ethiopia. This indicates not significant differences (H. Gankiso, (2017). In contrary Belyneh *et al* (2019) reported that the available phosphorous content of the soil between conserved and non-conserved plots did not have consistent pattern with conservation measures in Gumara watershed ,Northern Ethiopia. The application of Diammonium phosphate (DAP) may be the reason for its indistinguishable availability in the soil.

5.5. Cation exchange capacity (CEC)

The cation exchange is important in soils, as it measures the basic nutrient holding capacity (Bell 1993). Same study conducted by Bridge and Probert (1993) have noted that soils with low CEC(less than 10%) are poor at holding cationic nutrient. The CEC is a measure of the soil particles ability to exchange cations with freely mobile cations added to the soil matrix. The findings of Challa *et al.* (2016), Mengistu *et al.* (2016), and Selassie *et al.* (2015) reported SWC structures affect CEC of the soil. The conserved soil had significantly higher CEC than non-conserved soil. In Gumara watershed, Northwestern Ethiopia, Conservation measures caused a relatively higher CEC in conserved soils than in non-conserved but the difference did not show statistical significance. Sinore *et al.* (2018) reported a significantly higher CEC and exchangeable bases in a soil treated with susbania and elephant grasses than in controlled soil in Lemo District, Southern Ethiopia. Supporting terracing with such plants/ grasses strengthens the bund, generates high biomass, and increases OM and better control of erosion, consequently increases CEC in the soil. In Lole watershed, North Western Ethiopia, CEC showed a statistically significant difference ($p \leq 0.05$) between the treated and untreated fields. Soils in the treated fields showed significantly higher CEC than the untreated fields (M.Guadie et al 2019). In southern Ethiopia Gola micro watershed, the mean value of treated watershed 41.03 and untreated farm fields are 41.01 and the result reveal that CEC ranges from 34.7 to 49. 84 for treated and from 38.84 to 42.57 for untreated. It was not significant differences among treated and untreated fields. The value of CEC slightly increased in all slope classes. However, it was slightly decreased in untreated fields (H. Gankiso 2017). It is necessary to note that the higher levels of the CEC are mostly related to the higher level of organic carbon in the soil. These higher CEC, are the more cations it can retain.

Table 2 Details of studies on the impacts of SWC practices on soil physical and chemical properties. Positive impact (+), negative impact (−) and not studied (NS)

S.N	Studied structure	The place where the research conducted	Author	Impact on chemical properties	Impact on physical properties
1	Terraced hill side and farm land	Middle-Sulluh valley,Tigray	Solomon H. et al,2017	+	+
2	Level soil bund and level stone bund	Dawuro Zone SNNPR	Kebede et al,2019	- (SOC,AP,AK,PH)	(NS)
3	Conserved land with different structures	Gojjam Zone ,North E	Mengie et al,2019	+	(NS)
4	Physical soil and water conservation structures and integrated Physical soil and water conservation structures	Wolaita Zone, SNNPR	Tesfaye and Fanuel, 2019	+	+
5	Soil bund and stone bund	North Western Ethiopia	Mulat G.et al,2019	+	+
6	Soil bund	Arsi Negele	Dulo et al,2017	+	+
7	Stone bund	Dawuro Zone, SNNPR	Kero and Fanuel	+	+
8	Conserved land with different structures	Gojeb-river catchment,SNNPR	Melku et al,2020	+	+

Table 2 Selected Soil physico-chemical properties as influenced by conservation practices in cultivated lands of Bashe micro-watershed,2018

Conservation practices	BD (g cm−3)	Sand (%)	Silt (%)	Clay (%)	Textural class	pH-H2O	Soil OC (%)	Soil-OM (%)	Ava. P (mg kg−1)
NC	1.10	22.66	32.00	45.33	Clay	5.87	1.34	2.32	8.06
PSWC_2	1.07	24.00	29.00	47.00	Clay	6.07	1.68	2.90	11.74
ISWC-2	1.06	21.33	34.33	44.33	Clay	6.39	1.42	2.45	14.57
PSWC-5	1.06	21.33	40.33	38.33	C.loam	6.33	1.61	2.78	20.80
ISWC-5	0.96	23.00	32.33	44.66	Clay	6.60	1.74	3.00	25.23
CV (%	3.75	17.24	15.75	14.66	–	10.68	18.95	18.89	78.12

CV coefficient of variation, NC Non-conserved, PSWC-2 Physical SWC-2 year,ISWC-2 Integrated SWC-2 year,PSWC-5 Physical SWC-5 year,ISWC-5Integrated SWC-5 year,C.loam clay loam

Source - Tesfaye and Fanuel   (2019)

In the above table (table 1)   integrated SWC for 5 years reduced the soil bulk density; and increased soil pH (5.87 to 6.60), organic carbon (1.34 to 1.74%) and available phosphorous (8.06 to 25.23 mg kg−1) compared to non-conserved land. It showed that   SWC practices have positive impacts on soil properties of cultivated lands; however, they have more pronounced effect when physical SWC practices are integrated with biological SWC practices and at a longer establishment

Table 3- The mean and their significant variations (one-way ANOVA) of soil chemical properties in conserved and non-conserved plots

		pH (H2O)	SOC (%)	SOM (%)	TN (%)	Av. P (ppm)	CEC and Exch. cations [cmol(+) kg−1]				
							CEC	Na$^+$	K$^+$	Ca^{2+}	Mg^{2+}
Treatmentn	CL	5.77	2.49	4.3	.270	6.96	33.6	.31	.52	19.3	8.68
	NCL	5.66	1.66	2.86	.138	7.9	31.9	.18	.46	21.4	5.87
	F ratio	.772	4.457	4.457	8.504	.353	.195	12.36	.361	.425	9.515
	P	.389[ns]	.046[*]	.046[*]	.008[**]	.558[ns]	.663[ns]	.002[**]	.554[ns]	.521[ns]	.005[**]

Av. P available phosphorous, CEC cation exchange capacity, CL conserved land, NCL non-conserved land, ns not significant at p<0.05, Ppvalue, SOC soil organic carbon, SOM soil organic matter; **,* significantly different at p<0.01 and p<0.05 respectively (two-tailed); TN total nitrogen

Source - Belayneh *et al.* (2019)

Soil and water conservation practices have resulted in a statistically significantly higher mean values of total nitrogen, exchangeable Na+ and Mg2+ at p<0.01, and of soil organic carbon and organic matter at p<0.05 in the watershed(Table 2). The clay content, soil reaction, cation exchange capacity, and exchangeable K+ showed non-significant, but higher mean values in conserved plots. Furthermore, the effects of conservation practices on soil properties were found more effective in cultivated land uses as compared to that of grazing land uses. This is because conservation treatments had significant effects on organic carbon, total nitrogen, exchangeable Na+ and Mg2+ in cultivated land uses but only on exchangeable Na+ in grazing land uses. The interaction effect of treatments and land uses did not reach a statistically significant result for any of the soil properties considered in this study.

6.   Conclusion

Soil erosion is the deterioration of soil by the physical movement of soil particles from a given site. It is a natural process, which is usually accelerated by human action. It becomes a problem when human activity causes it to occur much faster than under normal conditions. SWC practices are the best option to sustainable agriculture, crop productivity and to increasing soil fertility. Terraces/bunds were important for many years especially during cropping season conserve the soil from runoff. SWC measures like terraces and bunds reduce soil erosion if they are correctly constructed and maintained, and if they are compatible with local environmental conditions. Conservation measures have important implications for improving soil fertility in many part of Ethiopia.

This review has shown that most physical SWC practices have positive impact on improving many soil physicochemical properties (soil texture, bulk density, TN, Av-P, SOC, pH, and CEC)when compared to non-treated land ; however, their effect is more pronounced when physical SWC practices are integrated with biological SWC practices .

Constructing physical structure alone does not restore the soil fertility to the level that the crop is demanding. To pronounce their effect physical SWC practices should integrate with biological SWC practices and for a longer establishment. Therefore, proper guidance and follow-up, use of agro-forestry and grass strips, and maintenance are required for the watershed's sustainability and good soil conditions. Supporting SWC structures with grasses and trees is very important for strengthening their effectiveness in improving soil fertility and decrease soil erosion in the watershed. Hence, integral use of both physical and biological SWC options and agronomic interventions would have paramount importance in improving soil quality.

7. Reference

Abay, C.; Abdu, A.; Tefera, M. Effects of graded stone bunds on selected soil properties in the central highlands of Ethiopia. Int. J. Nat. Resour. Ecol. Manag. 2016, 1, 42–50.

Ademe, Y.; Kebede, T.; Mulatu, A.; Shafi, T. Evaluation of the effectiveness of soil and water conservation practices on improving selected soil properties in Wonago district, Southern Ethiopia. J. Soil Sci. Environ. Manag. 2017, 8, 70–79. [CrossRef]

Adimassu, Z., Kessler, A., Hengsdijk, H., 2012. Exploring determinants of farmers' investments in land management in the Central Rift Valley of Ethiopia. Applied Geography, 35, 191-198.

Adimassu Z, Mekonnen K, Yirga C, Kessler A (2014) Effect of soil bunds on run-off, soil and nutrient losses, and crop yield in the central Highlands of Ethiopia. Land Degrad Dev 25:554–564

Alemayehu, M., Yohannes, F. and Dubale, P. (2006) Effects of Indigenous stone bunding (Kab) on crop yield at Mesobit-gedeba, north Shoa, Ethiopia. Land degradation and Development, 17, 45-54. doi:10.1002/ldr.693

Assefa A. Impact of terrace development and management on soil properties in Anieniarea west Gojjim. Master's thesis, Addis Ababa University, Ethiopia. 2007.

Berhe W (1996) Twenty years of soil and water conservation in Ethiopia: a personal overview. Regional soil conservation unit/ SIDA, Nairobi, Kenya

Berry, L., 2003. Land degradation in Ethiopia: Its extent and impact. Commissioned by the GM with WB support, 2-7.

Beshah T (2003) Understanding farmers: explaining soil and water conservation in Konso, Wolaita and Wello, Ethiopia. Wageningen University, The Netherlands, Doctoral Thesis

Challa A, Abdelkadir A, Mengistu T (2016) Effects of graded stone bunds on selected soil properties in the central highlands of Ethiopia. Int J Nat Resour Ecol Manage 1(2):42–50. https://doi.org/10.11648/j.ijnrem.20160102.15

Dejene, T. The Effectiveness of Stone Bund to Maintain Soil Physical and Chemical Properties: The Case of Weday Watershed, East Hararge, Ethiopia. Civil Environ. Res. 2017, 9, 9–16.

Demelash M, Stahr K (2010) Assessment of integrated soil and water conservation measures on key soil properties in South Gonder, NorthWestern highlands of Ethiopia. J Soil Sci Environ Manag 1(7):164–176

Desta, L., 2000. Land degradation and strategies for sustainable development in the Ethiopian highlands: Amhara Region. ILRI (aka ILCA and ILRAD).

Dubale, P., 2001. Soil and water resources and degradation factors affecting productivity in Ethiopian highland agro-ecosystems. Northeast African Studies, 8, 27-51.

Environment for Development (2010) Green accounting puts price on Ethiopian soil erosion and deforestation. Retrieved from http://efdinitiative.org/ourwork/policy-interactions/green-accounting-puts-price-ethiopian-soil-erosionand-deforestation

Fitsum S. Contribution of fanyajuu and normal bund for crop production in Ethiopia Addis Ababa, Ethiopia. 2002;83-84.

Gebiresilassie Y, Amare T, Terefe A, Yitaferaru B, Wolfgramm B, Hurni H (2013) Soil properties and crop yields along the terraces and toposequences of Anjeni watershed, central high lands of Ethiopia. J Agric Sci 5(2):134–144

Gemechu A (2016) Estimation of soil loss using revised universal soil loss equation and determinants of soil loss in Tiro Afeta and Dedo districts of Jimma zone, Oromiya National Regional State, Ethiopia. Trends Agri Econ 9(1–3):1–12. https://doi.org/10.3923/tae.2016.1.12

Gilligan DO, Hoddinott J, Taffesse AS (2009) The impact of Ethiopia's productive safety net programme and its linkages. J Dev Stud 45(10):1684–1706

Hana Gankiso 2017. Effect of soil erosion on physic-chemical properties of soils and crope yield:Goala micro watershed of Oyda wereda ,South Ethiopia

Haregeweyn N, Berhe A, Tsunekawa A, Tsubo M, Meshesha DT (2012) Integrated watershed management as an effective approach to curb land degradation: a case study of Enabered watershed, northern Ethiopia. Environ Manag 50(6):1219–1233. https://doi.org/10.1007/s00267012-9952-0

Hurni, H., 1993. Land degradation, famine and land resources scenarios in Ethiopia. In Pimentel D (ed.): World soil erosion and conservation. Cambridge University Press. DOI: 10.1017/CBO9780511735394.004.

Holden ST, Shiferaw B, Pender J (2001) Market imperfections and land productivity in the Ethiopian Highlands. J Agr Econ 52:62–79

International Monetary Fund (2005) Ethiopia: scaling up. Assessing the impact of a dramatic increase in aid in one of Africa's poorest countries. Finance Dev 42 Retrieved from https://www.imf.org/external/pubs/ft/ fandd/2005/09/andrews.htm

Kebede W (2015) Evaluating watershed management activities of campaign work in Southern Nations, Nationalities and Peoples' Regional State of Ethiopia. Environ Syst Res 4:6. https ://doi.org/10.1186/s4006 8-015-0029

Lal, R. (1996) Deforestation and land-use effects on soil degradation and rehabilitation in western Nigeria. III. Runoff, soil erosion and nutrient loss. Land Degradation &

Development, 7, 99-119. doi:10.1002/(SICI)1099-145X(199606)7:2<99::AID-LD R220>3.0.CO;2-F

Lowery, B., Hart, G.L., Bradford, J.M., Kung, K.-J.S. and Huang, C. (1999) Erosion impacts on soil quality and properties and model estimates of leaching potentials. In: Lal, R., Ed., Soil Quality and Soil Erosion, Soil and Water Conservation Society, Ankeny.

Ludi, E. (2004) Economic analysis of soil conservation: Case studies from the Highlands of Amhara region, Ethiopia. African Studies Series A18. Geographical Ber- nensia, Bernee.

Mengie Belayneh , Teshome Yirgu and Dereje Tsegaye.Effects of soil and water conservation practices on soil physicochemical properties in Gumara watershed, Upper Blue Nile Basin, Ethiopia 2019

Mitiku, H., Herweg, K., Stillhardt, B., 2006. Sustainable land management: a new approach to soil and water conservation in Ethiopia. Centre for Development and Environment (CDE) and NCCR North-South, University of Bern, Switzerland. DOI: 10.7892/boris.19217.

Mihrete, G. Effect of Soil Conservation Measures on Some Physico-Chemical Properties of Soil and Crop Yield in Simada District, South Gondar Zone, Ethiopia. Master's Thesis, Haramaya University, Dire Dawa, Ethiopia, 2014.

Molla T, Sisheber B (2017) Estimating soil erosion risk and evaluating erosion control measures for soil and water conservation planning at Koga watershed in the highlands of Ethiopia. Solid Earth 8:13–25. https://doi.org/1 0.5194/se-8-13-2017

MOA. Ministry of agriculture, Annual report Retrieved on December 2015

Mulugeta, D.; Karl, S. Assessment of integrated soil and water conservation measures on key soil properties in South Gonder, northwestern highlands of Ethiopia. J. Soil Sci. Environ. Manag. 2010, 7, 164–176.

Mulat Guadie , Eyayu Molla , Mulatie Mekonnen and Artemi Cerdà .Effects of Soil Bund and Stone-Faced Soil Bund on Soil Physicochemical Properties and Crop Yield Under Rain-Fed Conditions of Northwest Ethiopia 2019

Norton, D., Shainberg, I., Cihacek, L. and Edwards, J.H. (1999) Erosion and soil chemical properties. In: Lal, R., Ed., Soil Quality and Soil Erosion, Soil and Water Conservation Society, Ankeny.

Nyssen J, Poesen J, Moeyersons J, Deckers J, Haile M, Lang A (2004) Human impact on the environment in the Ethiopian and Eritrean highlands—a state of the art. Earth Sci Rev 64(3):273–320. https://doi.org/10.1016/S0012-8252(03)00078-3

Nyssen J, Poesen J, Gebremichael D, Vancampenhout K, Daes M, Yihdego G, Govers G, Leirs H, Moeyersons J, Naudts J, Haregeweyn N, Haile M, Deckers J (2007)

Interdisciplinary on-site evaluation of stone bunds to control soil erosion on cropland in Northern Ethiopia. Soil Till Res 94:151–163

Olson, K.R., Mokma, D.L., Lal, R., Schumacher, T.E. and Lindstrom, M.J. (1999) Erosion impacts on crop yield for selected soils of the North central United States. In: Lal, R., Ed., Soil Quality and Soil Erosion, Soil and Water Conservation Society, Ankeny.

Rahmato D (1994) Littering the landscape: environment and environmental policy in Wollo (northern Ethiopia). Paper presented at the International Symposium on African Savannas

Shiferaw, B., Holden, S.T., 1998. Resource degradation and adoption of land conservation technologies in the Ethiopian highlands: a case study in Andit Tid, North Shewa. Agricultural economics, 18, 233-247.

Sinore T, Kissi E, Aticho A (2018) The effects of biological soil conservation practices and community perception toward these practices in the Lemo District of Southern Ethiopia. Int Soil Water Conserv Res 6:123–130. https:// doi.org/10.1016/j.iswcr.2018.01.004

Solomon Hishe,James Lyimo and Woldeamlak Bewket   Soil and water conservation effects on soil properties in Middle Silluh Valley ,Northern Ethiopia ,2017

Tesfaye Tanto and Fanuel Laekemariam .Impacts of soil and water conservation practices on soil property and wheat productivity in Southern Ethiopia 2019

Troeh, F.R., Hobbs, A.J. and Danahue, R.L. (1980) Soil and water conservation for productivity and environmental protection. Prentice-Hall, Englewood Cliffs.

Wagayehu, B. and Drake, L. (2003) Soil and water conservation decision behavior of subsistence farmers in the Eastern Highlands of Ethiopia: A case study of the Hunde-Lafto area. Ecological Economics, 46, 437-451. doi:10.1016/S0921-8009(03)00166-6

Young, A. (1997) Agroforestry for soil management. 2nd Edition. CAB International, Wallingford.